I0487248

CARL GUSTAV JUNG'S PSYCHOTHERAPY:

THE PSYCHOLOGICAL AND MYTHOLOGICAL METHODS

ISBN: **978-1-4477-4740-6**

Copyright © 2011 Andreas Sofroniou.

Copyright © 2011 **ANDREAS SOFRONIOU.**

ISBN: **978-1-4477-4740-6**

CONTENTS   PAGE: 2

1: THE PSYCHOLOGICAL AND MYTHOLOGICAL METHOD

1.1 Popularisation of Method

The Swiss psychiatrist, Carl Gustav Jung popularised the terms *'introvert and extravert'*, interpreted the deeper conscious levels in terms of mythology, and established psychotherapy as the treatment of disorders with extensive research in various psychological methods.

1.2 Studies and Experience

Professor Jung (1875-1961) studied medicine at Basel, and worked under Bleuler at the Bugholsli clinic at Zurich (1900-1907). He established the term *'complex'* in his early studies in *word association*, and his 1907 publication of *The Psychology of Dementia Praecox* led to his meeting Sigmund Freud in Vienna. He became Freud's leading collaborator and was elected President of the International Psychoanalytical Association from 1910 to 1914.

1.3 Critical of Freud

His independent researches made him increasingly critical of Freud's sexual definition of the *libido*. His publication of *The Psychology of the Unconscious* caused a break in 1913. From then onwards he developed his own theories, foremost among which were his description of psychological types (*'extraversion/introversion'* 1921).

1.4 Self –regulating System

His theory of *psychic energy* emphasised a final point of view as against a purely causal one. His discovery and exploration of the *'collective unconscious'*, with its *'archetypes'* was an impersonal substratum underlying the *'personal unconscious'*; the concept of the psyche as a *'self-regulating system'* expressing itself in the process of *'individualisation'*.

1.5 Symbolism

To this latter process Jung devoted most of his latter work, constantly enlarging the scope of his researches; to include the interpretation of the dreams and drawings of patients, the symbolism of religions, myths, historical antecedents as (e.g. alchemy), and even modern physics *('synchronicity')*.

1.6 Influences on Sciences

Thus, Jung's work has become of great importance for medicine, psychology, anthropology, religion, art, history, literature, etc.

1.7 Condensed Picture

This presentation of Jung's psychotherapy is intended to give a condensed picture and an introduction to his extensive publications and method of therapy. Above all, to wet the reader's appetite for further interest in Jung's own extraordinarily voluminous works. It is the author's opinion that it would be inappropriate to attempt a description of Jung's forty plus years of intensive research, in a few pages that this book can afford.

1.8 Short Task

In short, a practically impossible task. It must necessarily remain a sketch, which the author attempts to organise as simply and clearly as possible, but that must renounce going into profundities or details.

1.9 Author's Justification

If one needs a justification as to why I still believe strongly in the application of psychotherapy, the half a century of practical application to the human mental health and the psychological illnesses, this practical experience alone should suffice. What I am attempting to do in this book is to bring together a short explanation of how C G Jung tried to bring a new method of therapy to the world. In the main, I would like the content of this write-up to be as close to the original work of the Master, and for the reader to decide how such thoughts can be applied in their search for happiness. Should there still be doubts, read it as if it is another collection of thoughts by a philosopher.

1.10 Science not Philosophy

To consider Carl Gustav Jung as another Thinker/Philosopher is still an honour for a man who devoted his time in bringing forward his philosophical thoughts. Lest not ignore the fact that the Jungian psychotherapists, and Jung himself consider their method to be science; neither a school of philosophy, nor a religion.

1.11 Conceptual Studies

It was in 1960 when I came across the concepts and application of psychotherapy, psycho-analysis, psychiatry, and mental health in general. Having previously spent four years in a general hospital, I had two years in which to finalise my thesis for my Doctorate in Psychology. I had a selection of cases to assess and report on successes and the failures of the medical treatment of mental patients as prescribed at the time.

1.12 Cane Hill Psychiatric Hospital

It all happened in the boundaries of five square miles on a hill in Surrey, U.K. where the psychiatric hospital kept 2,500 patients within its walls and another 2,000 people involved in giving various types of services. My critique on such methods of treatment made me biased toward the use of medical-less treatment and the best available type of psychotherapy. At that time it was that of the Jungian Psychotherapy and Freud's psycho-analysis.

1.13 Sufferings Demolished

Fifty years later I visited the site and there was no psychiatric hospital. The construction of suburbia houses was taking place instead. The sufferings of people were demolished together with the buildings and the methods of the somewhat cruel way of treating people. I returned to the warmth of my Wiltshire home, reminisced a little more and sat down to write.

2 JUNG'S PSYCHOLOGY

2.1 Term Established

Carl Gustav Jung established the term of Psychotherapy as a part of the wider aspect of his psychological and psychiatric studies. In this manner, the Jung Psychotherapeutic works are divisible into a theoretical part, whose principal headings can be described quite generally as:

- Nature and Structure of the Psyche,

- Laws of the Psychic Processes and Forces,

- The practical part based on these theories, their application, as therapeutic method in the narrower sense.

2.2 Philosophical Derivation

If one would reach a correct comprehension of Jung's *system*, one must first of all accept Jung's standpoint and recognise with him the full reality of the *psychic functions*. This point of view was, remarkable as it may sound, relatively new at his time. For up to a few decades earlier, the *psyche* was not considered as independent and subject to its own laws, but was studied and interpreted through derivation from philosophy, religion or from natural science, so that its true nature could not rightly be discerned.

2.3 Psychic Equals Physical

To C G Jung the psychic is no less real than the physical. Though it is not immediately touchable and visible, it is still fully and unambiguously experienceable. Even in the twenty-first century, it is a world in itself – subject to law, structured, and possessed of its special means of expression. All that we know of the world comes to us, as does all knowledge of our own being, through the medium of the *psychic*, which is therefore one of the most important aspects and conditions of experience.

2.4 Interchangeable Subjects

To study it as such was Jung's aim; not however to elevate it as would a mere psychologism to be the sole ground of all knowledge. The psychological, physical, and physico-mathematical standpoints (as well as many others) are interchangeable and can be studied at will according to the problem and the special interests of the enquirer.

2.5 Psychological Aspect

Jung took the psychological aspect, leaving the others to persons competent in their fields, drawing however upon his wide acquaintance with psychic reality, so that this theoretical structure is no abstract system created by the speculative intellect but an erection upon the solid ground of experience and resting only on that.

2.6 Principles

Its two main pillars were:

- The principle of *psychic totality*,

- The principle of *psychic dynamics*.

These two points were elaborated, together with the practical application of *the system*, in Jung's researches and book publications.

3 PSYCHE'S NATURE AND STRUCTURE

3.1 Psyche, Soul, or Mind

By using the term *'psyche'* Jung understood not merely what we usually mean by the word *'soul'* or *'mind'*, but the totality of all psychological processes, both conscious and unconscious. That is something broader than and including the soul, which for him constituted only certain limited complex of *functions*. According to his definitions, the *psyche* consists of two spheres supplementing one another but opposed in their properties – of *consciousness* and the so called *unconscious*.

3.2 Ego

The *ego* has a share of both. The following diagram shows the ego standing between two spheres, which not only supplement but also complement or compensate each other. That is, the dividing line that marks them off from each other in our ego can be displaced in both directions, as is suggested by the arrows and the dotted lines in the figure.

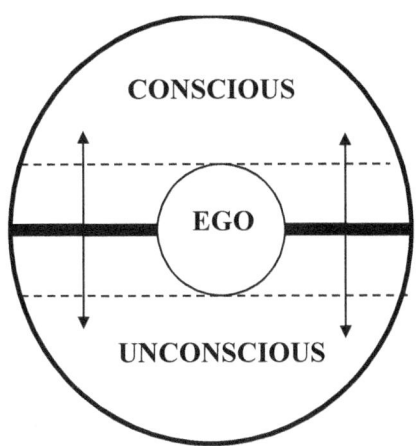

3.3 Centre of Reference

The ego itself is not exclusively *conscious*, but is conceived as a centre of reference for conscious and unconscious *psychic contents* alike. It forms, as a concept embracing the unitary totality of our *psychosomatic* beings. It is naturally only expedient of thought and an abstraction that the ego stands exactly in the middle.

3.4 Consciousness

Jung defines consciousness as *"the function or activity which maintains the relation of the psychic contents to the ego"*. The next diagram (paragraph 3.7) shows how the *sphere of consciousness* is surrounded by contents lying in the unconscious. Here are those contents which have been put aside (for our consciousness can take only a very few contents at once) but which can be raised again at any time into consciousness; furthermore, those which can be repressed because they can be disagreeable for various reasons - i.e., *"forgotten, repressed, subliminally perceived, thought, and felt matter of any kind."*

3.5 Personal Unconscious

This region Jung called it the *'Personal Unconscious'* in order to distinguish it from that of the *'Collective Unconscious'*, as is indicated in the following diagrammatic representation (see 3.7). For the *collective part of the unconscious* no longer includes contents that are specific for the *individual ego* and result from the personal acquisitions, but such as result *"from the inherited possibility of psychical functioning in general, namely from the inherited brain structure."* This inheritance is common to all humanity, perhaps even to the entire animal world, and forms the basis of every individual psyche.

3.6 Primal Datum

Further on, Jung maintained that the *unconscious* is older than *consciousness*. He added that it is the primal datum out of which *'ever afresh arises'*. Thus, consciousness is merely built upon the *fundamental psychic activity*, which consists in the functioning of the unconscious.

3.7 Unconscious Sphere

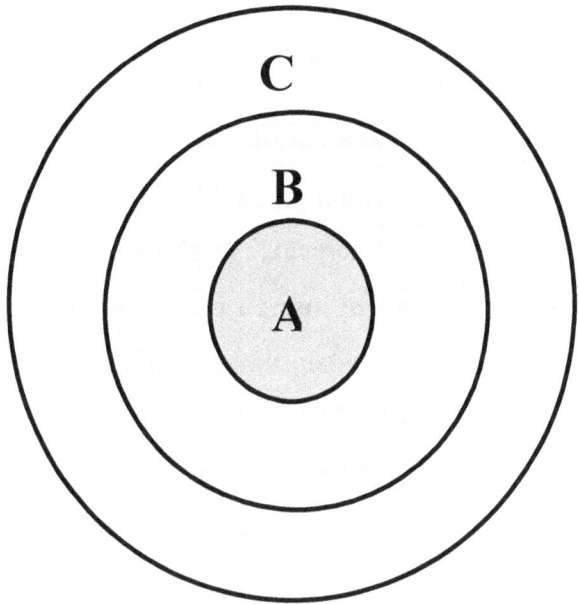

A: The part of the collective unconscious that can never be raised into consciousness,

B: The sphere of the collective unconscious,

C: The sphere of the personal unconscious.

3.8 Consciousness Dependent on Unconscious

The notion that man's psychic life is in the main conscious is false, for we spend the greater part of our life in the unconscious: we sleep or daydream... It is incontestable that every important situation in life our consciousness is dependent upon the unconscious. Jung added that *children begin life in an unconscious state and grow into a conscious one.*

3.9 Unconscious Contents

The unconscious consists of contents which are entirely undifferentiated, representing the precipitate of humanity's typical forms of reaction since the earliest beginnings, apart from historical, ethnological, racial, or other differentiations, in situations of general human character. For example, such situations as those of fear, danger, struggle against superior force, the relations of the sexes, of children to parents, to the father and mother imago, of reaction to hate and love, to birth and death, to the power of the bright and principles, etc.

3.10 Capacity of Unconscious

A basic capacity of the unconscious is that of acting compensatively and of setting up in contrast to consciousness, which normally always gives an individual reaction, adapted to outward reality, to the situation in question, a typical reaction derived from general human experience and conforming to *internal laws*, thereby making possible an adequate adjustment based on the totality of the *psyche*.

3.11 Psychic Totality

The diagram below serves as an illustration of the psychology and structure of consciousness. The circle symbolises the *totality of the psyche*; at the four points of the compass stand the *four basic functions* that are constitutionally present in every individual: *thinking, feeling, intuition, and sensation.*

3.12 Psychic Functions

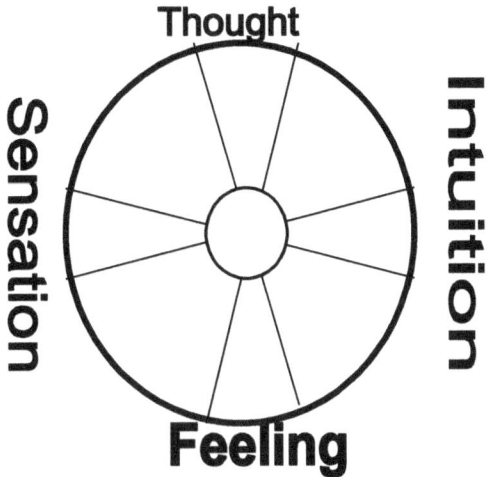

By a psychological function Jung understood a *"certain form of activity that remains theoretically the same under varying circumstances and is completely independent of its momentary contents."*

3.13 Decisive Facts

The decisive fact is not what one thinks, but that one employs one's *intellectual function* and not one's *intuition* in receiving and working up contents presented from without or within. *Thinking* is that function which seeks to reach an understanding of the world and an adjustment to it by means of an act of thought, or cognition, i.e., of conceptual relations and logical deductions. In contrast thereto, the *feeling function* apprehends the world on the basis of an evaluation by means of the concepts, pleasant or unpleasant, adience or avoidance.

3.14 Rational Functions

Both functions are characterised as rational because they work with values: thinking evaluates by means of cognitions from the viewpoint 'true/false, feeling by means of emotions from the viewpoint *'agreeable/disagreeable'*. These two fundamental forms of reactions are mutually exclusive as practical determinants of behaviour; the one or the other predominates.

3.15 Irrational Functions

The other two functions, *sensation and intuition*, Jung called the *irrational functions*, since they circumvent the ratio and work not with judgements but with mere perceptions, without evaluation or interpretation. *Sensation* perceives things as they are and not otherwise. It is the sense of reality par excellence, what the French call the *'fonction du rèel'*.

3.16 Inner Perception

Intuition perceives likewise, but less through the *conscious apparatus of the senses* than through its capacity for an unconscious *'inner perception'* of the potentialities in things.

3.17 Historical Event

The *sensation type* will take notice of an historical event in all its details but disregard the psychological context in which it is set; the *intuitive*, on the contrary, will pass over the details carelessly but perceive without difficulty the inner meaning of the occurrence, its possible relations and consequences.

3.18 Dominant Function

Although man possesses constitutionally all four functions, experience shows that it is always only one of these functions with which he orientates himself and adjusts himself to reality. This function becomes the dominant function for adjustment; it gives the conscious attitude its direction and quality and stands constantly at the disposal of the individual conscious will.

3.19 Typology

If we wish to give a complete schematic representation of the personality according to Jung's *typological system*, we can think of *introversion-extraversion* as constituting a third axis perpendicular to the cross axes of the four functional types. Referring each of the four functions to both the attitudinal types, we get an eightfold spatial figure. The idea of the quaternity is in fact seldom expressed by the double four, the eight, as well as by the four itself.

3.20 Mixed Types

The four functional types, based on the predominance of the one or the other function in the individual are valid in this form only theoretically. In real life they almost never occur pure but more or less as mixed types as is suggested in the following diagram.

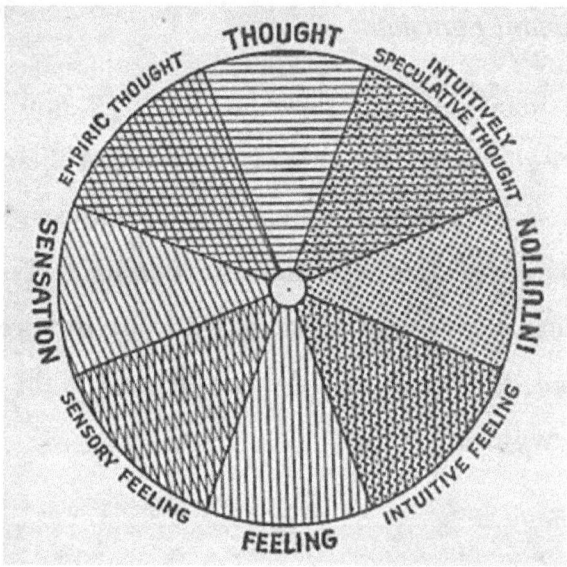

3.21 Over-differentiation

The complementary or compensatory relation of the functions to each other is a law inherent in the structure of the psyche. This almost inevitable *over-differentiation* of the superior function in the course of the years leads nearly always to tensions, which belong to the real problems of the second half of life and whose solution forms one of the principal tasks of this period.

3.22 Persona

This specific form of the general psychic behaviour of man with respect to the external world, which Jung called the Persona, is also connected with this over-differentiation.

3.23 Sphere of Consciousness

This diagram shows how the whole system of relations through which the psyche manifests itself in relation to the environment shuts off of the ego from the objective world.

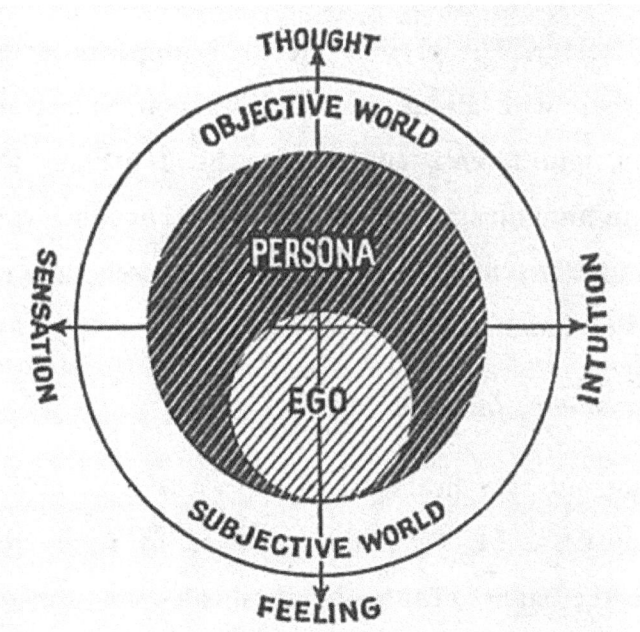

Sphere of Consciousness

3.24 Function-complex

Jung defines the *Persona* as follows: *"The persona is a function-complex which has come into existence for reasons of adaptation or necessary convenience, but by no means is it identical with the individuality. The function-complex of the persona is exclusively concerned with the relation to the object, to the exterior world. The persona is a compromise between the individual and society based on that which one appears to be."*

3.25 Stigmatisation

Adjustment to the environment can occasionally, however, be attempted not by means of the *superior function*, as is the rule, but by the *inferior*. It then succeeds correspondingly unsatisfactory. The persona inevitably appears in this case stigmatised with all the inadequacies that characterise the inferior, undifferentiated function. Such persons not only make an unnatural, artificial impression, but they can easily mislead the psychologically naïve to an entirely false estimate of their real nature.

3.26 Personality Inflation

But, not only the bearers and representatives of collective consciousness, the *'big names'* attested by community and society, the badges of title, dignities, roles, etc., constitute an attraction and therefore cause an *inflation of the personality*. Beyond our ego there is not just the collective consciousness of society but also the collective unconscious, our own deep, which conceals equally attractive and imposing figures.

3.27 Psychic Health

A well fitting and functional persona is an essential condition for the psychic health and is of the greatest importance if the demands of the environment are to be met successfully.

3.28 Character Index

The functional type to which one belongs would be in itself an index to a man's psychological character. It alone, however, will not suffice. In addition his general psychological attitude, i.e., his way of reacting to what meets him from without or within, must be determined.

3.29 Extraversion and Introversion

Jung distinguishes two such attitudes: *Extraversion and Introversion.* They represent orientations that essentially condition all *psychic processes*, the *reaction habitus*, namely through which one's way of behaving, of subjectively experiencing, and even of compensating through the unconscious is given.

3.30 Habitus, the Central Switchboard

This habitus Jung calls *"the central switchboard, from which on the one hand external behaviour is regulated and on the other specific experience is formed."* Extraversion is characterised by a positive relation to the object, introversion rather by a negative. The extravert follows in his adjustment and response patterns more to the external, collectively valid norms, the ideals of the time, etc. the introvert's reaction, on the contrary, is mainly determined by subjective factors.

3.31 Unsuccessful Adjustment

Thence comes his so often unsuccessful adjustment to the external world. The extravert thinks feels and acts in reference to the object; he displaces his interest from the subject out upon the object, he orientates himself predominantly by what lies outside him.

3.32 Orientation of Value

With the introvert the subject is the starting point of his orientation and the object is accorded at most a secondary, indirect value. This type of man draws back in the first moment in a given situation, as if with an unvoiced *'No'*, and only then follows his real reaction.

3.33 Empirical Material

Whereas the functional type describes the way in which the *empirical material* is specifically grasped and formed, the attitudinal type introversion-extraversion characterises the general psychological orientation, i.e., the direction of that general *psychological energy* which Jung conceives the *libido* to be. It is anchored in our biological constitution and is much more firmly determined from birth than is our functional type.

3.34 Conscious Effort

For, although, the choice of the superior or principal function is in general determined by a certain constitutional predisposition to the differentiation of a particular function, this latter can be greatly modified by conscious effort or thought or even repressed.

3.35 Inner Rebuilding

This is very seldom the case with a basic attitude or manner of reaction. Here only an *'inner rebuilding'*, an alteration in the psyche's structure, can bring about such a change, either through a spontaneous transformation (in this case again biologically determined) in puberty or the climacteric years or through a toilsome process of psychic development such as an analysis.

3.36 Ancillary Functions

Therefore the differentiation of a second and third function, i.e., of the two ancillary functions, is relatively easier than that of the fourth, *inferior function*, for the latter is not only the furthest removed from the *principal function* and standing in sharpest contrast to it, but it also coincides with the still unlived, obscure attitudinal type. The introversion of the extraverted thinking type, for example, has not the tone of intuition or sensation but primarily that of feeling.

3.37 Compensatory Relation

Extraversion and introversion stand likewise in *compensatory relation* to each other. If consciousness is extraverts, the unconscious is introverted, and conversely. This fact is of decisive significance for psychological understanding.

3.38 Matrimonial Problems

The difference in types is thus the real psychological basis of matrimonial problems, difficulties between children and parents, friction in relations of friendship and business, even indeed of social and political differences. Everything of which one is unconscious in one's own psyche appears in such cases *projected* upon the object, and as long as one does not recognise the projected content in one's own self the object is made into a scapegoat, the ethical task would then be to recognise in one's self the opposed attitudinal habitus, which is structurally given in everyone.

3.39 Equilibrium

Through its conscious acceptance and development the individual would not only come into *equilibrium* himself but also understand his fellowmen better.

3.40 Opposition of Functions

The *opposition of the functions* and of the conscious and unconscious attitude is intensified into a *conflict* in the individual, as a rule, only towards the second half of life; indeed it is just that problem which indicates an alteration of his psychological situation in that portion of life.

3.41 Attitudinal Type

As the functional type, so too the *attitudinal type* to which a person belongs almost always remains unknown to him or is mistaken. It is very difficult in any case and it requires a lengthy psychological investigation to isolate it from the kaleidoscopic picture that the psyche presents to the observer.

3.42 Constant Forms of Reaction

Extraversion and introversion are, indeed, generally *constant forms of reaction* in the life of one and the same person, although they can replace each other at times. Certain phases in human life and even in the lives of peoples are characterised more than extraversion, others more by introversion.

3.43 Eight Different Psychological Types

Combining extraversion and introversion as general attitudinal habitus with the four functions, these result in all *eight different psychological types*:

- the extraverted thinking type, the introverted thinking type,

- the extraverted feeling type, the introverted feeling type, etc.;

3.44 Compass

These (the Eight Different Psychological Types) form a kind of compass, with which we can orientate ourselves concerning the structure of the psyche.

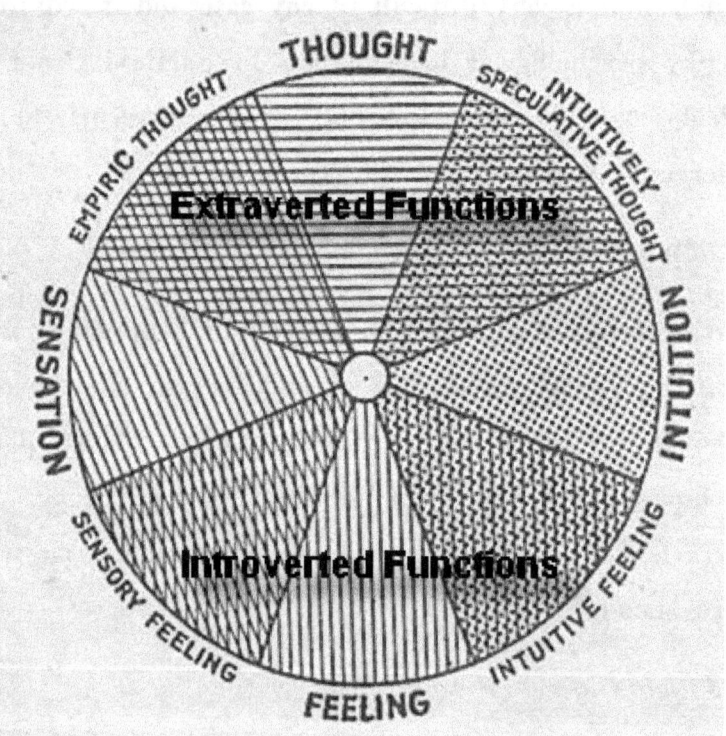

3.45 Sphere of Unconscious

As already mentioned, according to Jung's writings, the unconscious includes two regions, a personal, and a collective.

The next diagram gives a schematic representation of this classification.

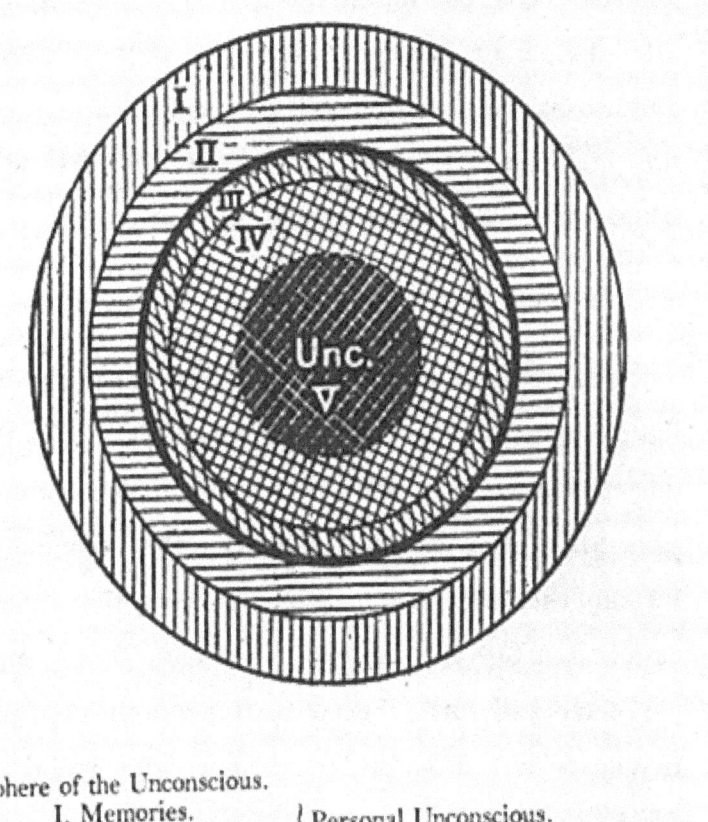

Sphere of the Unconscious.
    I. Memories.
    II. Repressed material. } Personal Unconscious.
   III. Emotions.
   IV. Irruptions from the deepest part of the un-
       conscious. } Collective
    V. That part of the unconscious that can never } Unconscious.
      be made conscious.

It has already been said what forms the content of the personal unconscious: forgotten, repressed, subliminally perceived thought, and felt matter of any kind.

3.46 Regional Divisions

But the collective unconscious is also divided into regions which lie over one another. The first, following downwards after the personal unconscious, is the region of our emotions and affects, the primitive drives, over which, however, when they manifest themselves, we can sometimes exercise control, which we can still somehow rationally order.

3.47 Autonomous Unconscious

The next region already includes those contents which break immediately out of the deepest, most obscure centre of our unconscious, never wholly to be made conscious, with elemental force, as foreign bodies that remain externally incomprehensible and never permit themselves to be assimilated fully by the ego. They have wholly autonomous character and form the contents not only of neuroses and psychoses but often of the visions and hallucinations of creative spirits.

3.48 Zones

To differentiate the various zones or their contents according to the zone to which they belong is often extremely difficult. They occur mostly in connection with each other, in a kind of mixture.

3.49 Symptom and Complex

The manifestations that first of all remain visible on the plane of consciousness are the *symptom and the complex*. The symptom can be defined as a phenomenon of the obstruction of the normal flow of energy and can manifest itself psychically or physically.

3.50 Broadening of Consciousness

It is a danger signal indicating that something essential in the conscious adjustment is disarranged or inadequate and that, accordingly, a *broadening of consciousness* ought to take place, i.e., a removal of the obstruction, although one is not always able to say in advance where the point of obstruction lies and how it is to be reached.

3.51 Arbitrary Functioning

Complexes Jung defines as *"psychological parts split off from the personality, groups of psychic contents isolated from consciousness, functioning arbitrarily and autonomously, leading thus a life of their own in the dark sphere of the unconscious, whence they can at every moment hinder or further conscious acts."*

3.52 Nuclear Element

The complex consists primarily of the *'nuclear element'*, which is mostly unconscious and autonomous and so beyond human influence, and secondarily of the numerous associations thereto, which in turn depend partly on the original personal disposition and partly upon experiences casually connected with the environment.

3.53 Ascending Complex

The following diagram shows the *ascending complex*, under whose thrust consciousness, as it were, is broken through and the unconscious, lifting itself over the threshold of consciousness, forces itself onto the conscious plane.

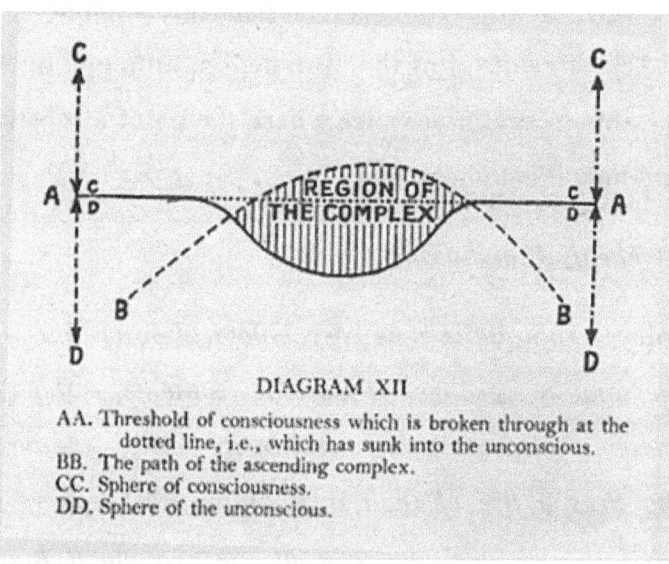

DIAGRAM XII

AA. Threshold of consciousness which is broken through at the dotted line, i.e., which has sunk into the unconscious.
BB. The path of the ascending complex.
CC. Sphere of consciousness.
DD. Sphere of the unconscious.

3.54 Passive State

The individual falls from an active, conscious state into a passive, *'possessed'* one. Such an ascending complex acts as a foreign body in the field of consciousness. It has its specific closed-ness, wholeness, and relatively high degree of autonomy.

3.55 Moral Conflict

It generally gives the picture of a disordered psychic situation, strongly toned emotionally and incompatible with the habitual conscious situation or attitude. One of its most frequent causes is, accordingly, moral conflict; by no means limited to the sexual. The conflict is a mental power before which at times the conscious will and the freedom of the ego cease.

3.56 Complexes

Everyone has complexes. All sorts of everyday slips, as Freud in his Psychopathology of Everyday Life has shown, testify to that unmistakeably. Complexes do not necessarily imply inferiority of the individual who has them; they merely indicate that something un-united, inassimilable, conflicting exists, perhaps a hindrance, may be too a stimulus to greater efforts and so even to fresh successes.

3.57 Psychic Points

Complexes are thus in this sense *focal and nodal points of psychic life* with which one would not wish to dispense, indeed on could not do without, for else psychic activity would come to a standstill.

3.58 Emotional Shock

The origin of the complex is frequently in a so-called *trauma*, an emotional shock, or the like, by which a fragment of the psyche is split-off. The complex probably has its ultimate basis as a rule, however, in the apparent impossibility of accepting the whole of one's own individual nature.

3.59 Association Method

The actual significance of a complex can only be demonstrated and the freeing of the individual from its influence accomplished by practical psychotherapy. Its presence, its effective depth, and its emotional tone can nevertheless be determined with the aid of the association method worked out by Jung, based on his researching methods.

3.60 Psyche Mechanism

The association method has proven that the *psyche mechanism* is able to point with clock-wise exactness to complex-laden points of the psyche. Jung has worked out and refined the association method to the utmost precision, in manifold detail, and from the most different points of view.

3.61 Didactic and Diagnostic Method

As a *didactic and diagnostic method* it has become an essential aid to psychotherapy and belongs today to the standard equipment of psychiatric institutions, clinical psychological training, and vocational guidance of every kind, and even finds its use in the law courts. The concept of the complex comes from Jung.

3.62 Dream

The easiest and most effective way of acquainting one's self with the mechanisms and contents of the unconscious is via the *dream,* whose material consists of conscious and unconscious, familiar and unfamiliar elements. These elements can occur in the most varied mixtures and can be derived from anywhere, beginning with the so-called *'remnants of the day'* and going on to the deepest contents of the unconscious.

3.63 Interpretation

Jung described their arrangement in the dream as standing outside of causality. Likewise space and time do not hold for them. The dream language is archaic, symbolic, pre-logical; a picture language whose meaning can only be discovered through special methods of interpretation.

3.64 Royal Pathway

Jung accords the dream extraordinary importance, regarding it not only as the way (the royal pathway) to the unconscious but as a function through which in great part the unconscious exhibits its regulative activity. For the dream gives expression to the other side, the one opposite to the conscious attitude. Unswayable by our consciousness, it is a pure manifestation of the unconscious, of that uninfluenced primal nature that Jung on this account calls the objective psychic.

3.65 Continuity of Processes

Consciousness aims always at the adjustment of the individual to the external world. The unconscious, on the contrary, is indifferent to this egocentric purposive-ness and partakes of the impersonal objectivity of nature, whose one goal is maintenance of *continuity of the psychic processes*; it is accordingly a guard against one-sidedness, which could lead to isolation, inhibition, or other pathogenic phenomena.

3.66 Standard Symbols

In view of the already mentioned highly significant compensatory function of the dream, which not only expresses fears and wishes but profoundly affects the whole psychic situation, Jung refused to set up *'standard symbols'*.

3.67 Manifold Contents

The contents of the unconscious are always manifold in meaning, and their significance depends equally upon the contents in which they occur and upon the specific external and internal situation of the dreamer. Many dreams even go beyond the personal problems of the individual dreamer and are the expression of problems that occur over and over again in human history and concern the whole human collective.

3.68 Prophetic

In this case, Jung professes that these types of dreams have *prophetic character* and are therefore regarded even today among primitives as the concern of the entire tribe and are publicly interpreted with great ceremony.

3.69 Fantasies and Visions

Besides dreams, Jung distinguishes also *fantasies and visions as bearers of the manifestations of the unconscious*. They are related to dreams and occur in states of diminished consciousness. They exhibit a manifest and a latent content, are derived from the personal or collective unconscious, and furnish thus material equivalent to that of the dream for psychological interpretation. From the ordinary wish fulfilling dream to the ecstatic vision, pregnant with meaning, their variability is unlimited.

3.70 Universal Human History

Themes of mythological nature, whose symbolism illustrates universal human history, and reactions of particularly intensive kind, allow one to surmise the involvement of the deepest layers.

3.71 Archetypes

These motives and symbols Jung named *Archetypes*. They are representations of instinctive, i.e., psychologically necessary responses to certain situations, which, circumventing consciousness, lead by virtue of their innate potentialities to behaviour corresponding to the psychological necessity, even though it may not always appear appropriate when rationally viewed from without.

3.72 Conscious Adjustment

It is this absolute inner order of the unconscious that forms our refuge and help in the accidents and commotions of life, if we only understand how to get in touch with it. So it becomes comprehensible that *the archetype can alter our conscious adjustment or even transform it.*

3.73 Augustine's Term

The archetypes, Jung has borrowed this term from Augustine, are akin to what Plato called the *'idea'*. Plato's idea may be understood only as a primordial image of highest perfection in its light aspect, aloof of earthly reality, whereas its dark counterpart does not belong to the world of eternity but to the ephemeral world of mankind.

3.74 Bipolar Structure

On the other hand, according to Jung's conception, the archetype is inherent in its *bipolar structure*; the dark side as well as the light. Jung also calls the archetypes the *'organs of the soul'*. They are only formally determined, not in regard to their contents; and their ultimate core of meaning may be delimited but never described.

3.75 Gestalt

If we wanted to look for further likely analogies the *'Gestalt'* in the broadest sense of this term, as used in Gestalt psychology and also in biology, should be mentioned in the first place.

3.76 Axial System

"The form of these archetypes," Jung said, *"is perhaps comparable to the axial system of a crystal, which predetermines the crystalline formation in the saturated solution, without itself possessing a material existence. This existence first manifests itself in the way the ions and then the molecules arrange themselves…"*

4 LAWS OF PSYCHIC PROCESSES AND OPERATIONS

4.1 Dynamic Movement

Jung conceives the total psychic system as being in continuous dynamic movement. By psychic energy he means to be understood the totality of that force which pulses through and combines one with another all the forms and activities of the psychic system. This psychic energy he called libido.

4.2 Libido

It is nothing else than the intensity of the psychic process, its psychological value, which is determinable only through its psychological manifestations and effects. The concept of libido is used here no differently from the analogous expression *'energy'* in physics – as an abstraction, that is, that expresses dynamic relations and rests upon a theoretical postulate confirmed by experience.

4.3 Equilibrium

The structure of the psyche is accordingly (for Jung) not statistically but dynamically constituted. As the building up and tearing down of cells keeps the physical organism in equilibrium, so the distribution of psychic energy determines the relations between the various psychic data, and all disturbances therein lead to pathological phenomena.

4.4 Finalistic Conception

The dynamic way of looking at events is a finally directed one, in contrast to the mechanic, which is casual. Yet this finalistic conception is not the only one, for Jung utilises all possible ways of looking at the problem. It is characteristic of his theory of dynamics, however, and is contained in its fundamental principle, the law of inevitable complementariness, according to which all psychological happenings must occur.

4.5 Enantiodromia

The problem of the opposites is for Jung *'law inherent in human nature'*. The psyche is a self-regulating system, and there is no equilibrium and no self-regulating system without opposition. Heraclitus discovered the most remarkable of all psychological laws, namely, the regulatory function of the opposites he called this *'enantiodromia'*, by which he meant that everything is turned into the opposite at one time or another.

4.6 Complementary or Compensatory Relation

All that has already been said by Jung, concerning the structure of the psyche: concerning functions, attitudes, relation of consciousness to the unconscious, of the dream to the waking state, etc., has been regarded from the point of view of this law of complementariness, according to which the various psychic factors stand in complementary or compensatory relation to each other.

4.7 Transposition

Only one must imagine this picture, transposed to the psychic system, to be very complicated, since one has to do here with an interconnected, closed system including in its turn many sub-systems of such communicating vessels. In this total system the quantity of energy is constant and only its distribution is variable.

4.8 Equivalent Replacement

The physical law of conservation of energy and the Platonic notion of the *'soul as that which moves itself'* are archetypically closely related. Plato also wrote that, *'no psychic value can vanish without being replaced by an equivalent'*.

4.9 Buddhistic and Primitive Theory

According to another ancient notion, *'the soul itself is this force; its conservation is implied by the idea of its immortality'*, and in the Buddhistic and primitive theory of transmigration is implied its *'unlimited capacity for undergoing transformation while being constantly conserved'*.

4.10 Law of Energy Conservation

According to Jung's law of energy conservation, when consciousness loses energy it goes over into the unconscious, activates its contents; archetypes, complexes, etc., which upon commence a life of their own and, breaking into consciousness, can cause disturbances, neuroses and psychoses.

4.11 Freedom of Intervention

In the psychic system it is the conscious that is able through its relative freedom of intervention to effect this reversal. Jung clearly states: *"it pertains to the creativeness of the psyche that interference in the mere natural order constitutes its very being. The creation of consciousness and the possibility of differentiating and broadening consciousness is its power to control and compel nature."*

4.12 Temporal Order

The dynamic movement is directed, and it is distinguished accordingly a progressive and regressive movement, in temporal order.

4.13 Co-ordinating of Opposites

The progressive movement is a process whose direction is given by consciousness and which consists in a continuous and unhindered development of the process of adjustment to the conscious demands of life and in the differentiation of the attitudinal and functional type necessary thereto. The adequate solution of conflicts and decisions of all kinds by taking into account, i.e., co-ordinating, the pairs of opposites is essential for this.

4.14 Regressive Movement

The regressive movement occurs when through failure of the conscious adjustment and the resulting intensification of the unconscious or through repression, etc., a one-sided but in its special nature unavoidable obstruction of energy is brought about, in consequence of which the contents of the unconscious become unduly charged with energy and swell upwards.

4.15 Regression

This can, in case of a partial regression, if consciousness does not interfere at the proper time, throw the individual back upon an earlier stage of development, from neuroses, or, if a total reversal takes place and the unconscious floods consciousness, lead to a psychosis.

4.16 Value Intensity

Besides the temporal succession, the movement of the dynamic process, and the libido moves not only forwards and backwards, progressively and regressively, but also inwards, progressively and regressively, but also inwards and outwards, corresponding to introversion and extraversion, the second important characteristic of this process is its value intensity.

4.17 Imagination

The specific from of manifestation of energy in the psyche is the image, brought up by the creative power of the imagination, the creative fantasy, out of the material of the unconscious, the objective-psychic.

4.18 Pictorialised Manifestations

This active, creative work of the psyche commutes the chaos of the unconscious contents into pictorialised manifestations, as they present themselves in dreams, in fantasies, in visions and, analogously to these, in every act creative art.

4.19 Control of Meaning

It determines ultimately the significance, corresponding to the value intensity, with which the images are laden, this significance, i.e., content of meaning, being measured by the constellation in which the image appears in the individual case.

4.20 Constellation

By constellation is understood here the setting of an image in a context according to which its value is determined. For in a dream, as an example, there are always a number of elements whose significance varies according to their positional value.

4.21 Correlation

The direction and intensity of the psychic dynamism correlate, they determine each other reciprocally; for the potential difference that is the primary condition of the process and direction of dynamic movement arises precisely from the difference in the energy charge, in the meaning present in the psychic contents.

4.22 Libido or Psychic Energy

The libido or psychic energy, as Jung conceives it, is the foundation and regulator of all psychic existence. This concept serves for the correct description of the actual processes in the psyche and their relations. It has nothing to do with the question of whatever or not there exists a specific psychic energy.

5 THE PRACTICAL APPLICATION OF JUNG'S THEORY

5.1 Relevant Fields

Jungian psychotherapy is not an analytical procedure in the usual meaning of this term, although it holds strictly to the medically, scientifically, and empirically confirmed premises of research in all relevant fields.

5.2 Healing and Salvation

It is a *'Heilsweg'*, a way of healing in both meanings of the German term, which signifies at the same time healing and salvation.

5.3 Removing Disturbances

Jung's method has all the requisites for healing a person from his psychic and therewith connected psychogenic sufferings. It has all the instruments for removing the most trifling psychic disturbances, the starting point of neurosis, and likewise for combating successfully the most complicated and threatening developments of mental disease.

5.4 Self-therapy

But besides this, it knows the way and has the means to lead the individual to his own healing, to that knowledge and perfection of his own personality which has ever been the aim and goal of all spiritual striving.

5.5 Complete Understanding

This way, is from its very nature, beyond all abstract exposition. Theoretic conceptions and explanations are adequate only up to a certain point for the comprehension of Jung's system of thought, for in order to understand it completely one must have experienced its vital working on one's self, or better put, *'undergone the method'*.

5.6 Psychological Guidance

The Jungian psychotherapy pertains thus, besides its medically effective aspect, an eminent capacity for psychological guidance, education, and the development of the personality. Both ways can but do not have to be followed at once. It probably follows from the nature of the matter that only a few are willing and determined to seek a way of healing, and these few take the way only out of inner compulsion, not to say necessity; *for this road is narrow as a knife-edge.*

5.7 Individual Case

For the endless variety of sufferings entrusted to his therapy Jung has set up no general prescription. The method applied and its intensity varies according to the circumstances of the individual case, to the psychic disposition and characteristics of the patient. Jung recognises the decisive role that sexuality and will to power play among men.

5.8 Freudian or Adlerian Points

Consequently there are numerous cases in which the illness is referable to disturbances in one of these driving factors and which therefore must be approached from a Freudian or Adlerian point of view. But while with Freud mainly the pleasure principle, with Adler the will to power acts as explanatory principle, Jung regards besides these other equally essential factors as motivating elements of the psyche and therefore rejects decisively the postulate that the predominate role in all psychic disorders belongs to one driving factor alone.

5.9 Spiritual and Religious

Besides these two assuredly significant ones there are for him still other highly important drives, before and above all that which belongs to man alone – the spiritual and religious need inborn in the psyche. This view of Jung's is an essential point in his theory, which distinguishes it from all other theories and determines its prospective-synthetic direction. For the spiritual appears in the psyche likewise as a drive, indeed as a true passion.

5.10 World of Drives

It is no derivative of another drive but a principle *sui generic*, namely, the indispensable formative power in the world of drives. Therewith Jung postulates from the first an equivalently ranking counter-pole to the world of the natural drives, of our primaeval biological nature, which forms, moulds, and develops this primitive nature and is peculiar to man alone.

5.11 Polymorphism

The polymorphism of primitive instinctive nature and the way of formation of personality confront each other as a pair of opposites called: nature and spirit. This pair of opposites is not merely the external expression but perhaps also the very basis of that tension which we call psychic energy.

5.12 Open System

We stand here at a decisive point that gives Jung's whole theory direction, tone, and depth and makes it an open system, excluding nothing of the stream of new problems that spontaneously follows all pioneer work in the world of the psyche. The attentive reader will believe that he finds conceptual contradictions in Jung's books. The science of the psyche must, nevertheless, set down the facts as it finds them, found not as an *either-or* but as an *either and or*. Thus this search for truth is at once cognition and encisioning.

5.13 Mystic

When the word *'mystic'* is uttered here more or less reproachfully, this only proves that people have forgotten that the strictest of the natural sciences, physics, is in its modern form neither more, nor less mystic than Jung's psychological systems, to which it exhibits the closest analogy of any of the natural sciences.

5.14 Dualism

One puts up here with what in the other case is called contradictory, with a real dualistic *'either and or'* that is forced to assert itself in the whole of contemporary physics, often only with the help of the boldest logical constructions, simply because reality compels it.

5.15 Modern Concepts

This dualism calls itself to our attention repeatedly in the formation of modern physical concepts, as when, for example, one must work with contradictory hypotheses concerning the nature of light (corpuscle or wave), or when all attempts to reconcile the field theory of reality with the quantum theory in logically irreproachable way fail.

5.16 Unity

Yet no one would therefore reproach the modern physicists with lack of logical skill and precision because the apparently illogical nature of the physical facts leads to a recognition of the irreconcilable, even of the paradoxical – naturally not without the hope and endeavour one day to win unity, even if not to force it.

5.17 Realm

The difficulty of psychology lies in the fact that, proceeding from and never leaving an empirical basis, it penetrates into the realm in which the expressions of language, derived from experience, are perforce inadequate and must remain a mere approximation.

5.18 Verification

Considered from this standpoint Jung is as far from being a metaphysician as any natural scientist ever was, for his statements always refer to empirically verified facts and are strictly limited to what is conceivable on the basis of experience. But here too, as in the modern natural sciences, experience leads us to a boundary where our empirical knowledge ceases and metaphysics begins.

5.19 Systematic Investigation

The domain of experience that he opened up and systematically investigated according to certain viewpoints in a scientific manner cannot by its very nature, however, be explored by the customary methods of the natural sciences, which postulate a purely conceptual treatment of their subject matter. Only the conceptually furthest advanced, because relatively simplest natural science, physics, has the possibility of clothing its bold hypotheses, unverifiable by any material constructions, in the pure, association-free language of mathematics.

5.20 Janus-head

Ultimately all modern psychology wears a *Janus-head*, a double face, one aspect of which is turned towards living experience, the other towards abstract cognition. It was not by chance that some of the greatest, most honest thinkers who lived in the conceptual and linguistic world of Europe, be it Pascal, Kierkegaard, or Jung, have to arrive, necessarily and fruitfully, at paradoxes when they occupied themselves with questions concerning no unambiguous matters but the ambiguous, two-faced nature of the psyche.

5.21 Synthesis

Jung's greatest step forward and the justification for the term *'synthesis'* is precisely his abandonment of the unambiguous casual thinking of the old psychology – namely, his recognition of the spirit must not be viewed as epiphenomenon, as sublimation, but as a principle *sui generic*, as a formative and therefore as the highest principle through which *Gestalt* organised structure, is psychologically and perhaps also physically possible.

5.22 Thinkers

For conclusions and thinkers such as Whitehead and Eddington have drawn from physics itself point actually to primary, formative, spiritual forces, which could be, and probably already have been, characterised as *'mystic'*.

5.23 Modern Logic

A remarkable identity in the form of expression employed by modern logic and Jungian psychology alike is perhaps also more than a coincidence, namely, the *'transcendence of problems'*, as both call it in the same words – where one has to do no longer with answerable questions but only with experienceable problems, with those problems which also form the content of Jungian psychological guidance and psychotherapeutic experience.

5.24 Freud, Adler, Jung

If one schematically compares the three principal tendencies in psychotherapy (Freud, Adler, Jung) with regard to the direction in which their central thought leads, one could say: Sigmund Freud looks for the *causae efficientes*, the causes of the later behavioural disturbances. Alfred Adler considers and treats the initial situation with regard to a *causa finalis* and both see in the drives the *causae materiales*.

5.25 Causae Formales

Jung, on the contrary, although he too naturally takes account of the *causae materiales* and like-wise takes the *causae finalis* as starting and end-point, adds to them something further and very important in the *causae formales*, those formative forces that are represented above all through the symbol and mediators between the unconscious and consciousness or between all the pairs of psychic opposites.

5.26 Reductive Method

Somewhat differently expressed this would mean: Freud employs a reductive method, Jung a prospective one. Freud treats the material analytically, resolving the present into the past, Jung synthetically, building up out of the actual situation towards the future, attempting to establish a relation between consciousness and the unconscious, between the pairs of the psychic opposites, in order to provide the personality with a basis on which a lasting psychological equilibrium can be built.

5.27 Dialectical Procedure

Jung's method is therefore not only to this extend a *'dialectical procedure'* in that it is a dialogue between two persons, and as such a reciprocal interplay of two psychic systems. It is in itself dialectic, as a process which, by confronting the contents of consciousness with those of the unconscious, i.e., those of the ego with those of the non-ego, calls forth a reaction between these two psychic realities that aims towards and results in bringing over both with a tertium quid, a synthesis.

5.28 Anonymity Abandonment

It is accordingly, too, from the therapeutic standpoint a preliminary condition that the therapists accept this dialectic principle equally as binding. In the field of dialectic procedure the physician/psychotherapist must step out of his anonymity and give an account of himself, exactly as he demands of his patient.

5.29 Ethical Value

It is at once apparent that the personality of the physician/psychotherapist, its worth and strength, its ethical value, is of the greatest importance in such a situation. As in any kind of medical and psychological treatment, through its attitude it plays a much more important and active role in Jungian analysis than in the methods of other depth schools of psychology.

5.30 Psychic Constitution

Even more than elsewhere the statement is valid here: the psychotherapist can bring the lead only to that point which he himself has gained. It is likewise true though that no therapist, however pre-eminent and skilled in his art he be, can obtain more from his patient than the values already there, for no labour on the soul can widen the bounds of the inner personality beyond those that were given it. So the possibility of mental and spiritual development remains more or less limited and the goal to be attained conditioned by the psychic constitution of the subject.

5.31 Interpretation

The patient alone determines the interpretation to be given the material he brings. Only his individuality is decisive here; for he must have a vital feeling of assent, not a rational consent but a true experience. Whoever would avoid suggestion must therefore look upon a dream interpretation as invalid until the formula is found that wins the patient's agreement. Otherwise the next dream or the next vision inevitably brings up the same problem and keeps bringing it up until the patient has taken a new attitude as a result of his experience.

5.32 Suggestively Influence

The often hard objection that the therapist could suggestively influence the patient with his interpretation could therefore only be made by one who does not know the nature of the unconscious; for the possibility and danger of prejudicing the patient are greatly over-estimated.

5.33 Independent Unconscious

The objective-psychic, the unconscious is, as experience proves, in the highest degree independent. If this were not so it could not at all exercise its characteristic function, the compensation of consciousness. Consciousness can be trained as in psittacosis case, like a parrot, but not the unconscious. If physician/therapist or patient errs in his interpretation, the unconscious, working always autonomously, corrects them strictly and uncontradictably in time.

5.34 Dream

The principal instrument of the therapeutic method is for Jung too the dream, it being that psychic phenomenon which affords the easiest access to the contents of the unconscious and which is especially suited because of its compensatory function to clarify and explain inner relations.

5.35 Dream Analysis

For the problem of dream analysis stands and falls with the hypothesis of the unconscious; without this dream is a senseless conglomerate of crumbled fragments from the current day. In the same way as the dream Jung utilises the fantasies and visions of his patients. If, therefore, in what follows we speak only of the dream for the sake of simplicity, fantasies, and visions are thereby also understood.

5.36 Conflicts and Manifestations

The fundamental difference between the Jungian and the other analytical methods consists in the fact that Jung sees in these phenomena, namely dreams, etc., not only contents of personal conflicts but in many cases also manifestations of the collective unconscious, which, going beyond the individual conflicts, sets over against these the primordial experience of universal human problems.

5.37 Determinate Functioning

Theory and method of Jungian dream analysis can only be sketched here briefly. Jung explained: *"The dream cannot be interpreted with a psychology taken from consciousness. It is determinate functioning, independent of will and wish, of intention and conscious choice of goal. It is an unintentional happening, as everything in nature happens… it is on the whole probable that we continually dream, but consciousness makes while waking such a noise that we do not hear it. If we could succeed in keeping a continuous record we should see the whole follows a definite trend."*

5.38 Purposiveness

This implies that the dream is a natural psychic phenomenon, but of a peculiar, autonomous kind, with a purposiveness unknown to our consciousness. It has its own language and its own laws, which one cannot approach with the psychology of consciousness – as subject, as so to speak.

5.39 Myths and Legends

For one does not dream: one is dreamed. We undergo the dream, we are the objects. One could almost say: we are able in dreams to experience as if they were real the myths and legends that we read when waking, and that is something essentially different.

5.40 Dream Roots

The roots of the dream, as far as we can tell, lie partly in the conscious contents – impressions of the day before, remnants from the current day, partly in the constellated contents of the unconscious, which in turn can come from conscious contents or from spontaneous unconscious processes. These latter processes, betraying no reference to consciousness, can be derived from everywhere.

5.41 Prophetic and Historic

Their origins can be somatic, physical and psychological events in the environment, or events in the past and the future; in the latter instance we may think , e.g., of dreams that bring a long past historical occurrence to life or prophetically anticipate a future one.

5.42 Fragmentation

There are dreams that originally had a reference to consciousness but have lost it, as if it had never existed, and now produce completely incoherent, incomprehensible fragments, then again such as represent unconscious psychic contents of the individual without being recognised as such.

5.43 Mysterious Message

Jung describes the arrangement of the dream images as standing outside the categories of *time and space* and subject to no causality. The dream is a mysterious message from our own night-aspect. The dream is never a mere repetition of previous experiences or events – certain categories of shock-dreams excepted – not even when we believe we recognise it as such. It is always knit together or altered according to its end, even though often inconspicuously, but ever in a different way from that which would correspond to the ends of consciousness and causality.

5.44 Structural Similarity

Jung found that that most dreams show a certain structural similarity. He conceived even their structure quite differently from Freud, dividing it as follows in the manner of a classical drama:

1. Time, place *dramatis personae*, that is the beginning of the dream, which frequently indicates the place where the action of the dream occurs and the persons acting therein,

2. **Exposition, i.e., the statement of the dream problem. Here the content is displayed, so to speak, that forms the basis of the dream, the problem, the theme that is given form by the unconscious in the dream, to which the unconscious will now make its pronouncement,**

3. **Peripety, which forms the backbone of every dream, the weaving of the plot, the intensification of events to a crisis or to a transformation, which may also consist in a catastrophe,**

4. **Lysis, i.e., the solution, the result of a dream, its meaningful conclusion, in which it points to the needful compensation.**

5.45 Base

This rough scheme, according to which most dreams are built up, forms a suitable basis for the process on interpretation. Dreams exhibiting no lyses allow one to infer a fatal development in the dreamer's life.

5.46 Lyses

These are specific dreams and they must not be confused with those which the dreamer recalls only fragmentarily or reproduces incompletely and which therefore end without any lyses. For naturally every phase of a dream can seldom be deciphered at once. It often requires a careful search before its structure is wholly revealed.

5.47 Conditionalism

Jung has introduced the concept and method of conditionalism into dream interpretation, i.e., under conditions of such and such a kind, such and such dreams can occur. The decisive factor is thus always the situation in question with its contemporary, momentary conditions. The same problem, the same cause may have, according to the total context, a correspondingly different significance; from the viewpoint of conditionalism they can have many meanings, not just always the same on without regard to the situation and the variability of their forms of appearance.

4.48 Expanded Causality

Conditionalism is an expanded form of causality, it is a manifold interpretation of causal relations and constitutes thus an attempt to conceive strict causality by means of an interplay of conditions, to enlarge the simple significance of the relation between cause and effect by means of the manifold significance of the relations between effects. Causality in the general sense is not thereby destroyed but only accommodated to the many-sided living material, i.e., broadened and supplements.

5.49 Amplification

Jung utilises no free association but a procedure that he calls *'Amplification'*. At the time he thought that free association indeed led *'always to a complex, of which it is nevertheless uncertain whether it is precisely this one that constitutes the meaning of the dream...'*

5.50 Dream Enrichment

Amplification means therefore, in contrast to the Freudian method of *"reductio in primam figuram"*, not a causality connected chain of associations to be followed backward, but a broadening and enrichment of the dream content with all possible similar, analogous images. It is further distinguished from free association in that the associations are contributed not only by the patient or dreamer but also by the psychotherapist.

5.51 Purpose

Freud asks with his reduction, *"Why?"*; Jung asks in dream interpretation above all, *"To what purpose?"* What did the unconscious intend, what did it want to tell the dreamer when it sent him exactly this and no other dream?

Jung distinguishes two kinds or levels of interpretation:

1. That upon the subjective level, and

2. That upon the objective level.

5.52 Subjective Level

Interpretation on the subjective level treats the dream figures, and events symbolically, as reflections of internal psychic factors and of the internal psychic situation of the dreamer. The characters of the dream then represent psychic tendencies or functions of the dreamer and the dream situation, his attitude in reference to himself and to the given psychic reality. The dream so conceived, points to internal facts.

5.53 Objective Level

Interpretation on the objective level implies that the dream figures as such are to be understood concretely and not symbolically. They then represent the dreamer's attitude to the external facts or persons to which he stands in relation.

5.54 Projection

Everything in the unconscious is projected, i.e., it appears as property or behaviour of the object. Only through the act of self-recognition do the corresponding contents then become integrated with the subject, therewith released from the object, and recognised as psychic phenomena. The phenomenon of projection is an integral part of the mechanism of the unconscious. *"A projection is never made, it happens!"* Jung defines it as *"displacing a subjective process out into an object,"* in contradistinction to introjection, which consists in taking in an object into a subject.

5.55 Symbol

The content of a symbol can never be fully expressed rationally. It comes out of that *'between-world of subtle reality which can be adequately expressed through the symbol alone'*. An allegory in a sign for something, a synonymous expression for a known content; the symbol, however always implies in addition something inexpressible through language.

5.56 Background Processes

Jungian psychotherapists maintain that Freud is therefore mistaken in calling *"those contents of consciousness which allow one to guess their unconscious backgrounds"* symbols, for according to his theory these contents play *"merely the role of signs or symptoms of background processes."*

5.57 Individual Significance

Jungian psychotherapists urge their patients so emphatically not only to set down their *'inward pictures'* in speech or writing but also to reproduce them in the form of their original appearance, in which not only the content of the picture but also its colours and their distribution all have a particular individual significance.

5.58 Conflicts

As Freud and Adler do, so does Jung hold the making and keeping conscious of conflicts for the *'conditio sine qua non'* of therapeutic success. Jung, however, does not refer the conflicts to a single drive, but regards them as consequences of a disturbance of the harmony between all the factors of the total psyche – between such factors, that is, as belong to the structure of the personal and such as belong to the structure of the collective component of our psychic totality.

5.59 Immediate Significance

Another difference in principle consists in the fact that Jung sought to solve all conflicts from the point of view of their immediate significance and not from that of the significance they had at the moment of their origin, without considering whether that moment lies far in the past or not.

6 SUITABLE SOLUTIONS

6.1 Life and Age Condition

Every condition in life and every age requires a solution suitable to itself alone; and therefore a conflict has a correspondingly different function and significance for the individual in question, even though its origin remain always the same. The way in which a man of fifty has to solve his parent-complex is altogether different from that in which a man of twenty has to do so, although the conflict may have arisen in both cases from identical childhood experiences.

6.2 Teleological

Jung's method is teleological: his view always takes in the totality of the psyche, bringing even the most circumscribed conflict with this totality. In this psychic whole the unconscious has the role not merely of a catch-basin for repressed contents of consciousness; it is above all *"the eternally creative mother of this very consciousness."*

6.3 Mental Trick

It is no *'mental trick'*, as Adler says, but on the contrary, the *"primary and creative factor in man, the never failing source of all art and of all human productivity."* This attitude and conception make it possible for Jung to see in a neurosis not merely something negative – a troublesome sickness, but also a positive, healing factor, a force in the formation of the personality, a broadening and deepening of consciousness.

6.4 Personality Broadening

A neurosis can thus act as a cry for help, sent from a higher, inner authority in order to call our attention to the fact that we urgently need broadening of our personality and that we can reach it if we confront the neurosis correctly.

6.5 Traumatic Neuroses

Jung would by no means deny that there are also neuroses of traumatic origin which result essentially from decisive childhood experiences and which must then be treated accordingly, i.e., following the Freudian principles. In many cases Jung used this method, which is especially suited for the younger persons, in so far as they are traumatically caused.

6.6 Acute Neurosis

Energy blocked up by the one-sidedness of consciousness can lead of itself in the course of life to a more or less acute neurosis, as can likewise an unconscious state that that is not accommodated to the requirements of the environment. At any rate it seems that, in spite of all, only a few succumb to the fate of the neurotic.

6.7 Self-realisation

Perhaps they are indeed *"actually the superior persons, who whatever reason have remained too long upon an inadequate plane,"* which their nature could not stand. One must not, however, presume that any *'plan'* of the unconscious lies behind this. *"The impelling motive, so far as it is possible for us to conceive of one, seems to be essentially simply a drive to self-realisation. One could also speak of a delayed ripening of the personality."*

6.8 Personality Wholeness

Wholeness of the personality is attained when all the pairs of the opposites are differentiated, when the two parts of the total psyche, the conscious and the unconscious, are joined together and stand in a living relation to one another. In this case, the psychological potential difference, the undisturbed functioning of the psychic life, is guaranteed by the fact that the unconscious can never be made completely conscious and always possesses the greater store of energy.

6.9 Individualisation

Realising one's self is a moral decision and this lends strength to the process of becoming one's self, which Jung calls the way of individualisation, becoming an individual being, and in so far as we understand by individuality our innermost, final, incomparable uniqueness, becoming one's own self.

6.10 Specific Nature

Individualisation does not mean individualism in its narrow, egocentric form; for individualisation makes the person into the individual being that he actually is. He does not therewith become selfish, but simply fulfils his own specific nature, between which and egoism or exaggerated individualism there is a world of difference.

6.11 Intense Analysis

The process of individualisation is an intense analytical effort which concentrates, with strictest integrity and under the direction of consciousness, upon the internal psychological process, eases the tension in the pairs of opposites by means of highest activation of the contents of the unconscious, acquires a working knowledge of their structure, and leads through all the distresses of a psyche that has lost its equilibrium.

6.12 Archetypal Symbols

To describe the archetypal symbols of the individuation process in all the manifold forms in which they appear it would require a thorough knowledge and consideration of the different mythologies and the symbolic accounts of human history. Without this they cannot be described and explained in detail. In what follows, therefore, a brief sketch must suffice, presenting only those symbolic figures that are characteristic of the principal stages of the process.

6.13 Shadow

The first stage leads to the experience of the shadow, which symbolises our *'other aspect'*, our *'dark brother'*, who albeit invisibly, yet belongs inseparably to our totality. For, *"the living form needs deep shadows in order to appear plastic. Without the shadow it remains a flat illusion."* The meeting with the shadow often coincides with the making conscious of the functional and attitudinal type to which one belongs.

6.14 Dark Aspect

The undifferentiated function and the rudimentarily developed attitude are indeed our *'dark aspect'*, that collective-human primordial disposition in our nature that one rejects from moral , aesthetic or whatever grounds and keeps in suppression because it stands on contradiction to our conscious principles.

6.15 Mechanism of Projection

Condition by our mechanism of projection, this dark aspect appears, as everything of which we are unconscious does, transferred onto an object. Therefore, the other is always guilty, when one does not consciously recognise that the darkness is in ourselves.

6.16 Personification

The shadow is an archetypal figure that often appears even today personified in many forms in the conceptions of primitives. It forms a part of the individual, a kind of split-off part of his being which is nevertheless joined with him just like a shadow. Therefore it means sorcery to the primitive when someone treads upon his shadow, and its evil effects can be made good again only by a series of magical ceremonies.

6.17 Depths of Unconscious

In art, too, the shadow is a popular and frequently treated theme; for the artist's inspiration and choice of themes come from the depths of his unconscious. What the artist creates in this way affects again the unconscious of his public, wherein ultimately the secret of his effectiveness lies. It is the figures of the unconscious that rise in him and appeals powerfully to men, although they do not know whence their fascination comes.

6.18 Collective Appearance

According to whether it belongs to the realm of the ego or the personal unconscious or to that of the collective unconscious, the shadow has a personal or collective form of appearance. It can, therefore, just as well show itself as figure from our field of consciousness, as the person who represents our opposites.

6.19 Anima/Animus

The second stage of the individuation process is characterised by the meeting with the figure of the *'soul-image'*, named by Jung the *Anima* in the man, the *Animus* in the woman. The archetypal figure of the soul-image stands for the respective contra-sexual portion of the psyche, showing partly how our personal relation thereto is constituted, partly the precipitate of all human experience pertaining to the opposite sex. In other words it is the image of the other sex that we carry in us, both as individuals and as representatives of a species.

6.20 Eve/Adam

According to psychic law everything latent, un-experienced, undifferentiated in the psyche, everything that lies in the unconscious and therefore the man's *'Eve'* and the woman's *'Adam'* as well, is always projected. In consequence one experiences the elements of the opposite sex that are present in one's own psyche.

6.21 Functional Complex

The soul-image is a more or less firmly constituted functional complex, and the inability to distinguish one's self from it leads to such phenomena as those of the moody man; dominated by feminine drives, ruled by his emotions, or of the rationalising, animus-obsessed woman who always knows better and reacts in a masculine way, not instinctively.

6.22 Natural Measure

The character of our soul-image, the anima or animus of our dreams, is the natural measure of our internal psychological situation. It deserves very special consideration in the way of self-knowledge.

6.23 Mother Bearer

The first bearer of the soul-image is probably always the mother; later it is those women who excite the man's fancy, whether in a positive or negative sense. The release from the mother is one of the most important and most delicate problems in the realisation of personality. The primitives possess for this purpose a whole series of ceremonies, initiations to manhood, rites of rebirth, etc., in which the initiant receives such instructions as shall enable him to dispense with guardianship of the mother. Only after this can he be recognised as an adult in the tribe.

6.24 Acquaintanceship

The European, however, must gain *'acquaintanceship'* with his feminine or masculine psychological component through the process of making conscious this component in his own psyche. That the figure of the soul-image, the contra-sexual in one's own psyche, especially with the Occidental is so deeply repressed in the unconscious and accordingly plays a decisive and often troublesome role is in great part the fault of our patriarchically oriented culture. *"It is accounted a virtue for a man to repress feminine traits as far as possible, as it was accounted unbecoming for the woman to be mannish".*

6.25 Persona

The soul-image stands in direct relation to the character of an individual's persona. If the persona is intellectual, the soul-image is quite certainly sentimental. As the persona corresponds to the habitual external attitude of an individual, so do the animus and the anima correspond to his habitual internal attitude.

6.26 Mediating Function

We can regard the persona as the mediating function between the ego and the outer world and the soul-image as the corresponding mediating function between the ego and the inner world.

6.27 Clarification

The diagram below attempts to clarify what has been stated by the Jungian followers and predominantly Jung himself.

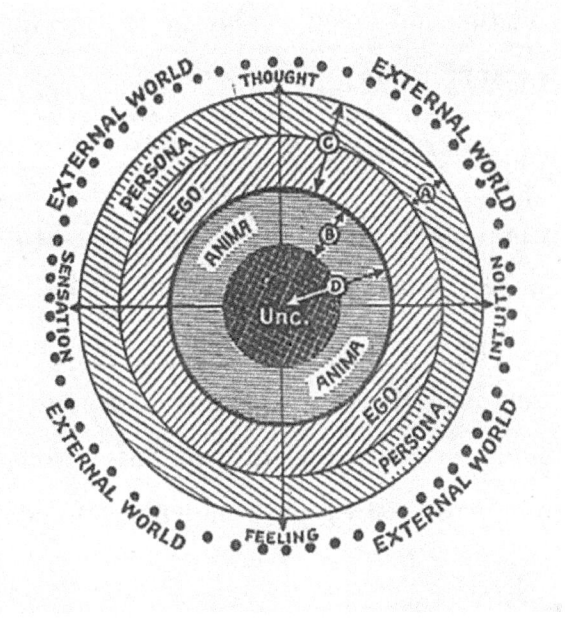

A. Would be the persona, lying as mediator between the ego and the external world,

B. The anima animus, represented as mediating function between the ego and the internal unconscious world.

C. Is at once ego and persona, representing the phenotypic, externally visible, manifest mental character,

D. Is the genotypic constituent, making up our invisible, latent, unconscious character.

6.28 Contra-sexuality

As the making conscious of the shadow makes possible the knowledge of other, dark aspect, so does the making conscious of the soul-image enable us to gain knowledge of the contra-sexual in our psyche.

6.29 Conscious Orientation

When this image is recognised and revealed, then it ceases to work from out of the unconscious and allows us finally to differentiate this contra-sexual component and to incorporate it into our conscious orientation, through which an extraordinary enrichment of the contents belonging to our consciousness and therewith a broadening of our personality is attained.

6.30 Unconscious Pictures

The following two plates are pictures from the unconscious that attempt to represent through symbolic figures an incorrect and a correct function of reference to the contra-sexual.

6.31 Heavy Burden

The first plate shows the conjuctio as it generally is but as it should not be. In the world of drives man and woman are here indivisibly one. Instead of striving together towards the sun of the spirit, they turn away from each other and sorrowfully carry it on backs as a heavy burden:

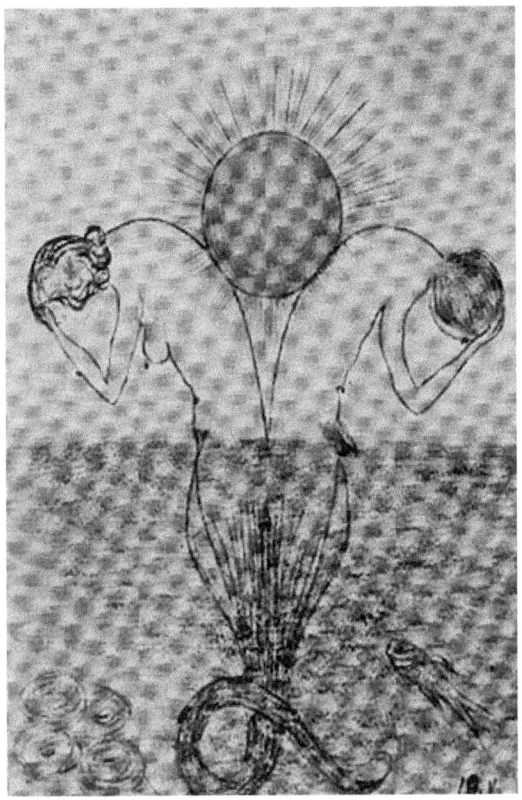

6.32 Creative Union

The next plate on the contrary shows the conjuctio as a true and creative union. The unconscious animal sides of man and woman are not grown together indivisibly but join one another in the symbol of the *'healing snake'*, which helps them to raise out of the depths of the sea the symbol of the unconscious, the *'precious stone'*, the symbol of the *Self*, without which the interlacing *Tree of Life* could never blossom.

6.33 Self-knowledge

According to Jung, the more on becomes conscious of one's self through self-knowledge and corresponding action, the more that layer of the personal overlying the collective unconscious vanishes. Thence arises a consciousness no longer captive in a petty and personally sensitive ego-world but participant in a wider, in the world of the objects.

6.34 Egoistic

This broader and deeper consciousness is also no more that sensitive, egoistic bundle of personal ambitions, wishes, fears, and hopes that must be compensated or perhaps corrected by unconscious personal counter-tendencies, but it is a function of reference connected with the object, the outer world, placing the individual in unconditional, binding and indissoluble community with it.

6.35 Objective Criterion

This renewal of the personality is a subjective state, whose real existence can be confirmed by no external criterion; therefore too is every further attempt at description and explanation useless, and only be who has had this experience is in a position to comprehend and attest its actuality. An objective criterion can be given for it just as little as, e.g., for happiness, which in spite of that possesses absolute reality.

6.36 Self

The *Self* is a magnitude super-ordinate to the conscious ego. It includes not only the conscious but also the unconscious portion of the psyche and is therefore a personality, so to speak, which we too are. We know that the unconscious processes stand in a compensatory relation to consciousness, not in a *'contrasting'*, because unconscious and consciousness are not necessarily opposed. They complement each other in the *Self*.

6.37 Psychic Totality

The next schema attempts to give a representation of the total psyche, placing the Self in the middle between consciousness and the unconscious, so that it has a share in both yet includes both in its rays; for the Self is not only the mid-point but also the circumference, taking in consciousness and the unconscious; it is the centre of the psychic totality, as the ego is the centre of consciousness.

6.38 Psychic System

The drawing is intended to show that the *Self* both forms the centre and includes the surrounds the whole psychic system with power of its radiation. The different parts of the total psyche are likewise included in the diagram, without any claim being made to represent their order, positional value, etc., it being impossible to show anything so abstract schematically. The only content of the Self that we really know is the ego.

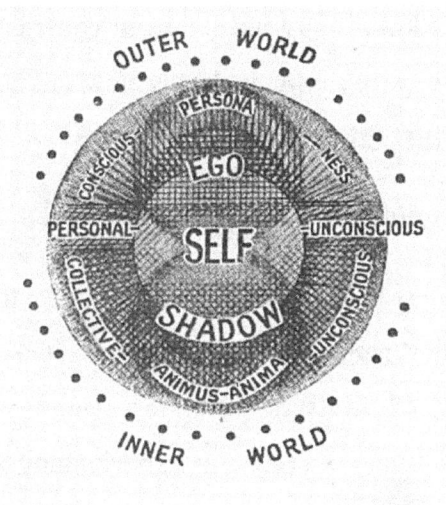

6.39 Psychic Category

The *Self* is also a psychic category, experienceable as such; and if we abandon psychological language we might name it the *'central fire'*, our individual share in God, or the *'little spark'*. Jung adds: *"It is that focal point of our psyche in which God's image shows itself most plainly and experience of which gives us the knowledge, as nothing else does, of the significance and nature of our likeness to God. It is the early Christian ideal of the Kingdom of God that is within you. It is the ultimate experienceable in and of the psyche"*.

6.40 Individuation Process

The individuation process, which could only be sketched briefly, consists in the gradual approach to the contents and functions of the psychic totality and in the recognition of their effect upon the ego. It brings one inevitably to acknowledge one's self for what one by nature is, in contrast to that which one would like to be, and probably nothing is more difficult for man than just this.

6.41 Self-realisation

Self-realisation is a way to give meaning, to form character, and thus to construct a higher consciousness. All consciousness of motives and intentions is germinating consciousness. Every increase in experiencing and knowledge means a further step in the development of one's personality.

6.42 Transformation Processes

So, Jung's psychology and the attempt to reveal the eternal processes of psychic transformation to the western man are only a step in the process of development of deeper human consciousness, which finds itself upon the way to unknown goals, and no metaphysic in the usual sense.

6.43 Intuitive Reality

First of all and thus far it is only psychology, but thus far also experienceable, understandable, and... real; an intuitive and therefore living reality. Jung's satisfaction with the psychologically experienceable and his rejection of the metaphysical are intended to imply no gesture of scepticism pointed against belief or faith in higher powers... Every pronouncement about the transcended should be avoided, for it is always only a ridiculous presumption of the human psyche unaware of its limitations.

6.44 Middle Way

To go the *'middle way'* is the task of the mature, for the individual's psychological situation is different at every age. At the beginning of life he must struggle out of infancy, which is still wholly imprisoned in the collective unconscious, to the differentiation and demarcation of his ego. He must get rooted in real life and, first of all, master the tasks; sexuality, profession, marriage, descendants, ties and connections of all kinds, that it imposes on him.

6.45 Superior Function

Therefore, it is of the greatest importance that he acquires the tools for his establishment and adjustment by means if the highest possible differentiation of his constitutionally superior functions. Only when this task, which constitutes that of the first half of life, is fully accomplished, should the experience of and adjustment to the internal be added to the adjustment to the external.

6.46 Attitude Completion

Once the construction and reinforcement of the personality's attitude with respect to the outer world are completed, energy can be turned to the as yet more or less unheeded inner psychic realities and can therewith bring human life to true perfection.

6.47 Middle Life Task

The establishment of the wholeness of the personality is a task of middle life. It seems to signify a preparation for death in the deepest sense of this world. For death is no less important than birth and, like the latter, belongs inseparably to life. Nature herself, if we only understand her aright, takes us here in her protective arms.

6.48 Continual Loss

The older we become, the more the outer world veils itself, losing continually in colour, tone, and attraction; and the inner world calls us and occupies us all the more. The aging individual nears ever more the state of dissolution in the collective psyche, out of which as a child he once with great effort he emerged.

6.49 Cycle Closure

So the *cycle of life* closes meaningfully and harmoniously, as has been expressed symbolically since the most ancient times in the pictures of the of the various tribes, the snake that bites its own tail. If this task is rightly fulfilled, then death must lose its terror and take its place meaningfully in the wholeness of life.

6.50 Infantile Adults

One must add the limitation, though, that apparently many do not even succeed in completing the task laid upon them by the first half of life, as the innumerable infantile adults prove, and that therefore life's rounding off through self-realisation is granted only to few. Just those few, nevertheless, have ever been the creators of culture, in contrast to those who have only produced and furthered civilisation.

6.51 Primal Nature

For civilisation is always a child of the ratio, the intellect; culture, on the other hand, arises out of the spirit, and the spirit is never bound to consciousness alone as is the intellect, but includes, forms, and controls at the same time all depths of the unconscious, the primal nature.

6.52 Over-differentiation

It is the particular and peculiar fate of western man, because historical conditions, origins, and spirit of the times are always determining factors also in the individual's psychological situation, that his instinctive side has withered through the over-differentiation of the intellect in the course of centuries, and that he has wholly lost the natural relation to his unconscious.

6.53 Turbulent Unconscious

For an explanation of the turbulent unconscious, Jung quotes: *"He has become so unsure of his instincts that he is tossed hither and thither like a floating reed on the swollen, turbulent sea of the unconscious, or as we have been able perturbedly to observe in the latest events, is already overwhelmed and swallowed up by the waves."*

6.54 Anchoring

Self-realisation is thus no fashionable experiment but the highest task that the individual can set himself. In regard to one's self, it means that possibility of anchoring one's self in that which is eternal and indestructible, in the primal nature of the objective-psychic. Thereby, the individual places himself again in the eternal stream, in which birth and death are only stations along the way and the meaning of life no longer lies in the ego.

6.55 Tolerance and Kindness

With regard to one's fellow-men, it summons up that tolerance and kindness in him which only he can give; who has searched out and consciously experienced his own darkest depths. With regard to the collective, its especial value consists in the fact that it is able to present to it that individual fully sensible of his responsibilities who from the personal experience of his psychic totality is aware of how the particular is obligated by its relation to the general.

7. THE JUNGIAN SYSTEM CLAIMS

7.1 Neither Religion nor Philosophy

The Jungian system of psychotherapy claims, in spite of its intimate reference to the fundamental problems of our being, to be neither philosophy, nor religion. Jung's followers insist that the *'system'* is the scientific summary and representation of all that the experienceable totality of the psyche includes; and as biology is the science of the living physical organism, so is the science of the living organism of the psyche.

7.2 Equipment

Thus it comprises also the whole of the equipment with which men have ever created and experienced religions and philosophies. It alone give the possibility of forming a well balanced living; that is not merely taken over traditionally and uncritically but that can be worked out and personally shaped by the individual with the help of these material and tools.

7.3 Reassurance and Comfort

No wonder that this system precisely today, when the collective psyche threatens to become all and the individual psyche nothing, is able to afford us reassurance and comfort; and that the task imposed by it, although it belongs to the most difficult of all times, lays it as an obligation upon us to bridge over the opposition between individual and collective through the full personality, standing in relation to both!

7.4 Reason over Instinctive Nature

The predominance which our reason, our one-sidedly differentiated intellect, has gained in the West over our instinctive nature and which expresses itself in our highly developed civilisation in a masterful technic that seems to have lost every connection with the external depth of the psyche, can only be compensated by calling to aid the creative powers lying there, restoring them to their rights, and elevating them to the heights of this intellect.

7.5 Transformation

This transformation, however, can only begin with the individual, for the masses are blind beasts. If this transformed individual has recognised himself as *God's likeness* in the deepest ethical sense of obligation, then (as Jung says) *"he will be on the one hand excellent in knowledge, on the other excellent in will, and no arrogant superman!"*

7.6 Culture of the Future

As a finale to this précis on the science of Jungian psychotherapy one can maintain that *the responsibility and the task of the culture of our future belong more than ever to the individual!*

BIBLIOGRAPHY:

PSYCHOLOGY OF THE UNCONSCIOUS. C G JUNG, 1913

DIE PSYCHOLOGIE VON C G JUNG. K W BASH, 1942

THE PSYCHOLOGY OF C G JUNG. JOLAN JACOBI, KEGAN PAUL, TRENCH, TRUBNER & CO LTD, 1943

INDIVIDUAL PSYCHOLOGY. ALFRED ADLER, 1914

AN OUTLINE OF PSYCHO-ANALYSIS. SIGMUND FREUD, THE HOGARTH PRESS, 1939

PSYCHO-ANALYSIS TRANSLATION. BY JAMES STRACHEY, 1949

THE INTERNATIONAL PSYCHO-ANALYTICAL LIBRARY. EDITED BY ERNEST JONES ,1949

THE PSYCHO-ANALYSIS OF CHILDREN. ANNA FREUD, 1942

CHILDREN'S PSYCHOANALYSIS. MELANIE KLEIN, 1949

TOTEM AND TABU. SIGMUND FREUD

STUDIEN ÜBER HYSTERIE. BREUER AND FREUD

BEYOND THE PLEASURE PRINCIPLE. SIGMUND FREUD

EGO AND I. SIGMUND FREUD

WHY WAR? SIGMUND FREUD AND ALBERT EINSTEIN, 1933.

THE PSYCHOPATHOLOGY OF EVERYDAY LIFE. SIGMUND FREUD, 1938

THE INTERPRETATION OF DREAMS. SIGMUND FREUD, 1936

THE CANE HILL EFFECT. ANDREAS SOFRONIIOU, ISBN: 978-1-4452-7636-6 (POETRY)

MORAL PHILOSOPHY, FROM HIPPOCRATES TO THE 21ST AEON.

ANDREAS SOFRONIIOU, ISBN: 978-1-84753-463-7

THERAPEUTIC PHILOSOPHY FOR THE INDIVIDUAL AND THE STATE.

ANDREAS SOFRONIIOU, ISBN: 978-1-4092-7586-2 & 0952795655

PHILOSOPHIC COUNSELLING FOR PEOPLE AND THEIR GOVERNMENTS.

ANDREAS SOFRONIIOU, ISBN: 978-1-4092-7400-1 & 0952795663

MORAL PHILOSOPHY, THE ETHICAL APPROACH THROUGH THE AGES.

ANDREAS SOFRONIIOU, ISBN: 978-1-4092-7703-3 & 0952725339

MEDICAL ETHICS THROUGH THE AGES. ANDREAS SOFRONIIOU, ISBN: 978-1-4092- 7468-1

PSYCHOLOGY OF CHILD CULTURE. ANDREAS SOFRONIIOU, ISBN: 978-1-4092-7619-7